YOUR KNOWLEDGE HAS VALUE

- We will publish your bachelor's and
  master's thesis, essays and papers

- Your own eBook and book -
  sold worldwide in all relevant shops

- Earn money with each sale

Upload your text at www.GRIN.com
and publish for free

Multi Purpose Seedling Need Assessment. The Case of Wolaita Zone, especially Humbo, Damot Woyde, Sodo Zuriya and Sodo Ketema

Zerihun Lemma

Bibliographic information published by the German National Library:

The German National Library lists this publication in the National Bibliography; detailed bibliographic data are available on the Internet at http://dnb.dnb.de.

ISBN: 9783346339652
This book is also available as an ebook.

WOLAITA SODO ATVET COLLEGE

A BASE LINE SURVEY ON:

MULTI PURPOSE SEEDLING NEED ASSESSMENT: *THE CASE OF WOLAITA ZONE; SEPECIALLY, HUMBO, DAMOT WOYDE, SODO ZURIYA AND SODO TOWEN*

BY:

ZERIHUN LEMMA

DEPARTMENT OF NATURAL RESOURCE

OCT, 2011E.C

WOLAITA SODDO, SOUTHERN ETHIOPIA

Table of Contents

CHAPTER ONE

1.1. Justification

Deforestation is a contributor to global warming and is often cited as one of the major causes of the enhanced greenhouse effect. According to the Intergovernmental Panel on Climate Change, deforestation mainly in tropical areas, could account for up to one-third of total anthropogenic carbon dioxide emissions. But recent calculations suggest that carbon dioxide emissions from deforestation and forest degradation contribute about 12% of total anthropogenic carbon dioxide emissions with a range from 6 to 17%. Plants remove carbon in the form of carbon dioxide from the atmosphere during the process of photosynthesis.

Ethiopia is one of the countries where forest resources degradation exhibits primarily due to population explosion. The forest policy is also not said to be effective in forest policy regards. At present it is facing a serious ecological problem triggered mainly by the fast growth of population which led to destructive nature of land resource uses involving deforestation even without leaving time space for regeneration of natural forests. The situation exposes the land to serious land degradation altering local environment and micro climate. To change this situation urgent natural resource management primarily forest management strategies intervention is required.

1.2. Statement of the Problem

There are a number of problems facing the forestry sub-sector. The forest resources have been declining both in size and quality. The high forests which used to cover 16% of the land area in the early 1950s were reduced to 3.6 % in the early 1980s and further declined to 2.7 % in the early 1990s Million, 2001. In Ethiopia the ecological crisis is deepening. It is deemed to be the result of misguided an unregulated modification of the Ethiopian environment, in particular the vegetation, soils and natural ecological processes. Increased human and animal population, whose livelihood is based on the use of natural resources, in particular renewable natural resources, has led to their fast depletion and serious degradation. Their exploitation has been and still is beyond their "self-replicating capacity"Tedla and Lema (1998).

In order to feed alarmingly growing population and mitigate adversely changing weather condition, it is mandatory to conducting pre survey around seedling need assessment in pilot areas and diverse the agricultural practices and natural resource management through plantation of multipurpose

1

trees, agro forestry, orchards, soil and water conservation through forage trees and grasses and at the same time developing quality feeds, securing seedlings at drought seasons. At present the need to multipurpose seedlings on farms is on the increase. It is difficult, however, for smallholders to access – at the right time, in the right quantities and of high quality the seedlings that they want to plant. Thus to achieve social and environmental needs of the community, there should be carried out effective base survey around seedling need assessment of community.

Therefore, this seedling need assessment is aimed to know accepting capacity of multipurpose seedlings of selected Woreda and to know their right time of seedling distribution

1.3. Assessment Objectives

1.3.1. General Objective

☐ To assess the need of multipurpose seedlings of selected worada's

1.3.2. Specific Objectives

✓   To assess the need of trees and coffee seedlings of the selected worada's
✓   To know the yearly accepting potential of seedlings of the selected area
✓   To assess the availability of Permanent tree nursery at study worada's
✓   To assess the perception of community towards the establishment of permanent seedling nursery in their vicinity

1.4. Significance of the multi-purse seedling need assessment

Zonal governments has established seedling nurseries in deferent worada's. But due to lack of conducting clear assessment on seedlings for pilot areas, a lot of multipurpose seedlings are left there and distributed unplanned areas. So as before raising deferent types of seedlings on the nursery bed, conducting pre seedling need assessment for pilot area is very crucial point. The aim of this study is to assess seedling need of the selected worada's. So this base line assessment study will indicate, which Woreda will need how much seedlings, when, how and it shall put out important solutions for seedling shortage problem in the area. Also the research will help as a base line for any researcher who are engage in similar study.

1.5. Scope of the study

The scope of this study was to assess the need of multipurpose seedlings of pilot worada's' and to assess the community perception towards different seedlings and its production areas. Especially the study limited in the Wolaita zone; Damot woyde woreda, Himo,Sodo zuriya and Sodo town.

CHAPTER TWO

2. Literature Review

2.1 Important concepts and Definitions of Multipurpose seedling nursery

In this Research multipurpose seedlings means a young plants that can grow from seed; it can includes tree seedlings, coffee seedlings, forage and other fruit seedlings having many benefits. Also the term multipurpose seedling is synonymies with agro forestry species.

2.1.1 Tree nursery

A **nursery:** is a place where seedlings are propagated, managed and grown to plantable size Vincent (2016). To ensure a good planting program me, good nursery stock is essential. Major causes of seedling mortality on-farm include the wrong size or poor health of the seedlings at the time of planting or poor health of the seedlings at the time of planting. Poor seedlings are likely to have slower growth, to be less able to compete with weeds or drought, and to be more liable to damage by insects and pests. Further, in a poor nursery, fewer seedlings will be raised from a given quantity of seed, and there will be considerable waste of money and time. After planting, the plants are immediately exposed to a harsh environment, and are more susceptible to damage from drought, grazing, fire and insects. Thus sound nursery practice is the foundation of any successful (on-farm and/or forestry) planting programme scheme.

2.1.2 Seed

A seed is a small embryonic plant enclosed in a covering called the seed coat, usually with some stored food. It is the product of the ripened ovule of gymnosperm and angiosperm plants, which occurs after fertilization and some growth within the mother plant. The formation of the seed completes the process of reproduction in seed plants (started with the development of flowers and pollination), with the embryo developed from the zygote and the seed coat from the integuments of the ovule. All seeds have different sizes, shapes and colours. Seeds of woody plants exhibit a great range of variation in shape, size, colour and behavior. The most essential factor for the success of plantation is the ready availability of quality seeds. The quality of seed is responsible for the future performance of each and every seedling.

2.1.3 Seedling

A seedling is a young plant that is grown from a seed. It is any young plant, especially one grown in a nursery for transplanting. Seedling development starts with germination of the seed (Joel Buyinza, 2016).

2.2. The status of forest resources in Ethiopia context

Ethiopia is recognized as an important origin and center of biodiversity and endemism on the African continent. This is generally related to physical and climatic diversity of the country. The long influence of an agriculture based culture and the increasing population size have put pressure on ecosystems and wildlife. The result, as one can imagine, has been the increase in the clearing of forests and vegetation for the purpose of mainly cultivable land and energy sources. In the past one hundred years, Ethiopia has faced a major loss of its forest cover from 35 to 2.5 and 3 percent of its landmass, which is 113 million hectares. This harsh reality indicates that an estimated land size of 163, 600 hectares of natural forest gets depleted annually. Most of the remaining forests are included in the National Forest Priority Areas that are under the management authority of regional states where the forests are situated. Currently, the Conservation and development of forests is regulated by Forest Conservation, Development and Utilization Proclamation No. 94 of 1994 (FDRE, 1994). The Proclamation recognizes three types of forests, namely; state and protected forests, regional and protected forests and private forests (Teshome, 2014).

2.3. Effects of Population Growth in Forest Resources

According to several reports Ethiopian forest covered approximately 40% of the land a century ago but now has shrunken to only 3% (Berry, 2003). Deforestation and consequent land degradation are global menaces, and so are they in Ethiopia. The forest degradation in Ethiopia is closely linked to the ongoing population growth. More people generally lead to an increasing demand on land for living and for agricultural production.

The situation got more severe in the eightieth when large numbers of people moved to South West Ethiopia in scope of organized resettlement programs. Consequently the pressure on the forest *resources* themselves increased due to a higher demand on fuel wood and construction timber. Finally, uncontrolled logging and the illegal export of wood stems to urban centers is a threat for the natural high forest of the country. The natural regeneration of the forest resources is difficult due to high populations of grazing and browsing livestock within the forests.

Forest resources such as poles, fuel woods, fodder, timber logs, medicinal valued plants, incense gum and resin, are the direct economic benefits that one can get it from forest ecosystem. Forest ecosystem also provide other environmental benefits that are non-quantifiable like, climate amelioration, watershed management, nutrient recycling, wildlife habitat provision, and others are being provided by forest ecosystem to human beings and other wild and domestic animals. Destruction of the forest ecosystem due to population pressure makes a human being out of these benefits.

CHAPTER THREE

3. Research Methodology and Description of the study Areas

The study will be conducted by taking four Woredas namely,Damot woyde,Sodo zuriya,Sodo town and Humbo woreda from Wolaita Zone as a case. Wolaita Zone (WZ) is the home of the people of Wolaita and is located in the Southern Nations, Nationalities, and Peoples Region (SNNPR), one of the regional states in the Federal Democratic Republic of Ethiopia. The Zone is bordered by Hadiya and Kembata Tembaro Zones in the north, Gamo Gofa Zone in the south, Dawro Zone in the west, and Sidama Zone and Oromia Regional State in the east (Wolaita Finannce and Economic Development,2018).The establishment of Wolaita 21 as a zone dates back to 2000 with the merger of 12 districts. Currently the projected number of population of Wolaita zone is 2018340. The major soil types found in the Zone are Nitosols, haplic Yermo soils, eutric Cambisoils, orthic Andisoils and calcaric Fluvisols. The topography of the Zone is featured by some rugged terrain with steep hills and mountains, flat plains and bush and shrub lands. As Desalegn states, altitude ranges from 1100 to 3000 meters above sea level. Rainfall occurs in two distinct rainy seasons, „kremt" rains (also called the „big rains") occurring in summer (roughly June, July and August) and „belg" rains (also called the „small rains") occurring in spring (roughly the mid-February to mid-May period). Kremt is the main production season, but the occurrence of rain during the belg season is equally important as it has implications on the food security of the households. In normal circumstances, rainfall ranging from 1100 to 1300 mm is expected in the higher altitudes while in the lowlands, it may average from 600 to 900 mm (WFED) 2018). Average temperature varies between 15 and 20 °C in the Zone.

MAP OF THE STUDY AREAS

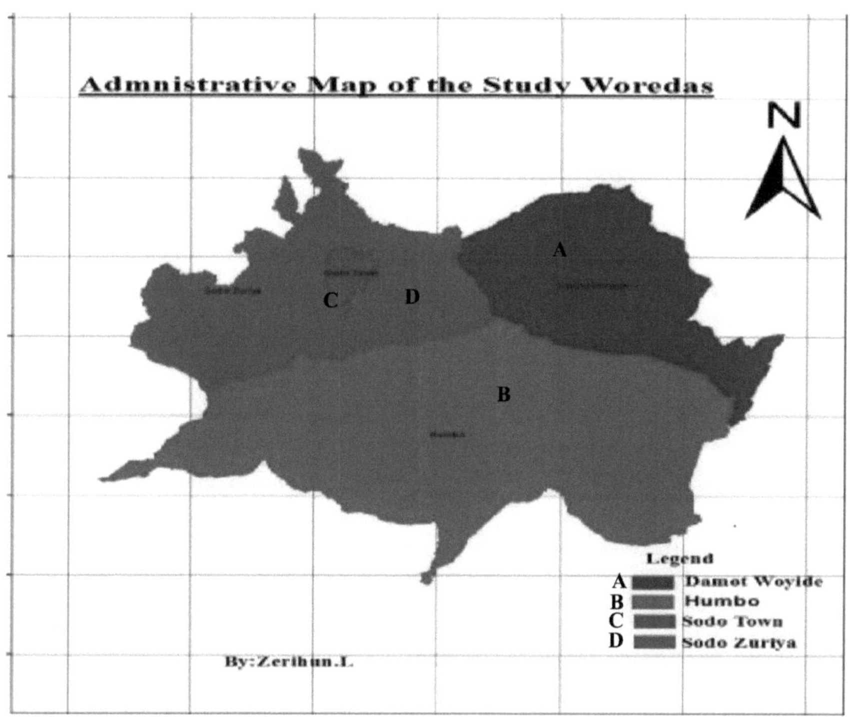

Figure 1Map of the study areas
Source: My own Computation

3.1 Research Strategy and Design

Research strategy is a way of systemizing observations, describing methods of collecting evidence and indicating the types of tools to be used during data collection (Cavaye, 1996). There are several types of research strategies e.g. case studies, surveys, experiments, ethnography, phenomenology, grounded theory, action research and archival analysis (Denscombe, 2007). Depending on the purpose of the research, more than one research strategy may be used. Every strategy provides an alternative way of collecting and analyzing empirical evidence. The choice of a research strategy is not accidental. It is informed by three main factors, namely the type of research questions posed, the extent of control the researcher has over the actual behavioral events and the degree of focus on contemporary events (Yin, 2003). Hence, the researcher has to make strategic decisions to put him- or herself in the best possible position to gain the best research outcome.

3.2 Research Design

In this survey, the descriptive research design was employed to make intensive investigation of the pre seedling need assessment of four worada's of Wolaita zone. Hence, to maintain triangulation in its analyses and findings, the design manifested the basic features of both the qualitative and quantitative researches.

3.3. Ways of Data Analyses and Presentation

The study areas Damot woyde, Humbo, Sodo Zuriya and Sodo town selected Purposive sampling approach (nonrandom sampling method).As it itemized above, their selection criteria clearly stated(They all have plenty of land; but those had not been rehabilitated, afforested, vegetated and green).

To carry out the study in all population of the study area, there were time, budget and labor limitations. Thus, (60HHs)was selected from the total of four worada's of the study areas" purposively. The 60 households were taken by giving equal proportion to each Woreda. From 60 households, 85% (51) were taken from male and 15% (9) were from female households. To minimize sampling biases, systematic random sampling methods were conducted. Finally, systematic random sampling method was employed by taking the nth element of the sample frame. A questionnaire, interview survey were used to collect the required data. Based on collected data the analysis was done using software like SPSS and Microsoft Excel and data w ere presented by different data presentation tools like tables and graphs.

CHAPTER FOUR

4. Data Analysis and Presentation

This chapter brings us to the results of findings and analysis on data gathered from the study areas. In doing so, the chapter has utilized different items and inputs; these are the results of the respondents" idea, voices of the informants and the relevant works of scholars relating to the issues under investigation. The chapter begun by providing a general description and analysis on the demographic characteristics of the respondents such as their Kebeles, sex, educational level, age composition of the sampled house-hold and socio economic characteristics of respondents those are house hold income, land size, of sampled households were analyzed. The other respective variables that related with the assessment of multipurpose seedlings, its availability of seedlings and nursery in the study area and perception of community towards shortages permanent nursery problems were analyzed.

4.1 Demographic and Socio economic Characteristics of Sampled Household

4.1.1 Ages of the respondents

Most of the sampled households in the study area, were ranges 35-45 age classes which indicates as, the most potential stages for any activity. The others were more than 45 aged households and finally, from the age27-37 were the most youth groups also those can capable any kind of works and jobs the area. The survey research revealed as most of the households in the study areas were, young group those can carry out any kind of works that related with natural resource management in particular and agricultural works in general. (See fig-2) below.

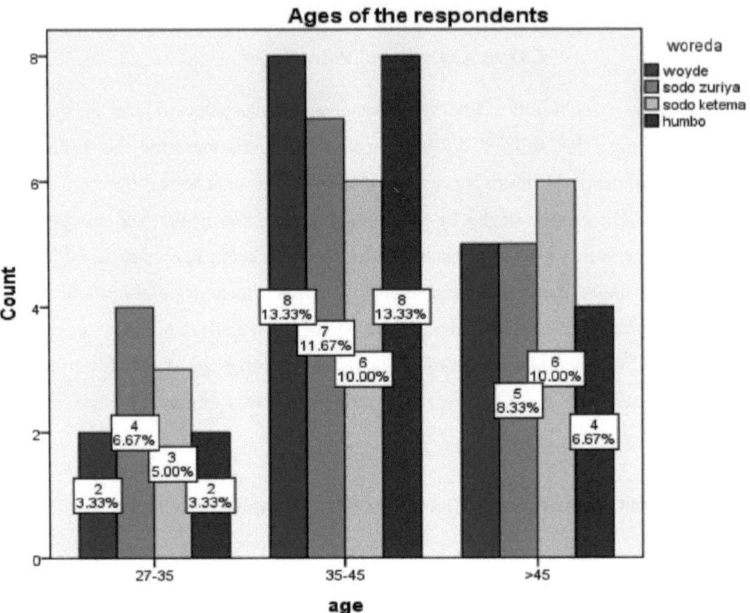

Figure 3 Ages of the respondents

Source: Survey results (2011)

4.1.2. The Educational Back Ground of the respondents

Another socio economic data of the area was educational level of the respondents. If there is availability of educated individuals in the given area, the area will not be challenged with the shortage of skill man power for a given industry. The abundance of educated human power in a given country considered as a pocket money for a country. The survey result shows about a greater than five percent of respondents were holds certificate and above. From them about 80%were dwellers of soddo town which indicates as the access of educational institution of an area. The left about 13% of respondents were completed 1-8and 10grade.In another word this indicates as most of the rural community completed primary school, but all of them can write; this is the most important variable for agricultural productivity and adoption and dissemination of a new technology towards them. See figure (3)

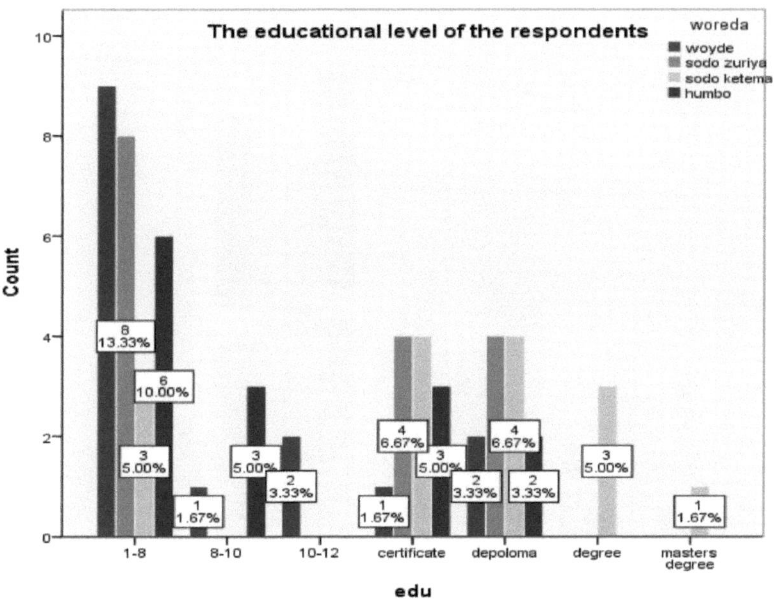

Figure 4 Educational level of the respondents

Source: Survey result (2011)

4.1.3. The land size of the households hold

Land is a fundamental asset for economic development, food security and poverty reduction in sub-Saharan Africa and has a crucial importance to the economies and societies of the region contributing a major share of GDP and employment, and constituting the main livelihood basis for a large portion of the population (Cotula, Toulmin et al. 2004). Likewise, land is a vital asset for a country like Ethiopia, where the country's economy is based on agriculture. As the base line of this empirical evidence, it has try to see the land size of each household in the study areas. As the survey assessment shows in all four study areas, the farmers holds 0.25ha-1ha (16%).On the other hand about 7%were having of0.25ha whereas the other household hold averagely about 5% which is less than 0.25ha. The survey result itemized as, most of the household in the area having 0.25ha and less than it. In this manner farmers cannot produce better products, which may result food insecurity of hall family and the family will going to fail under a harsh condition. see figure(4)

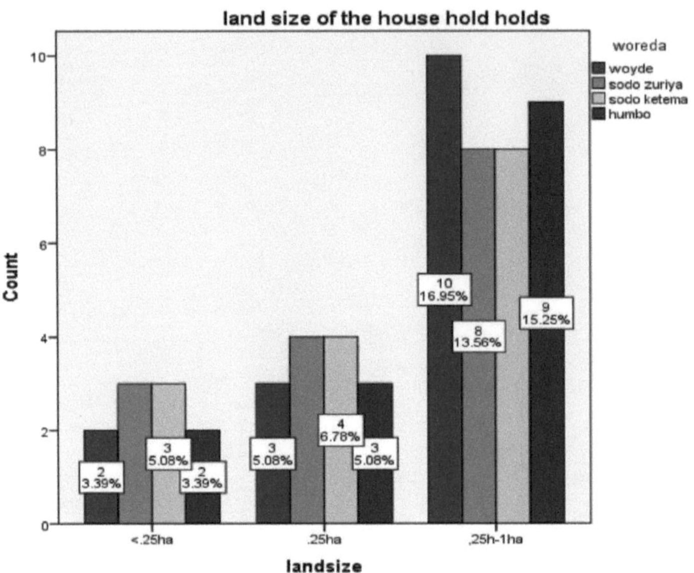

Figure 5 the land size of house hold holds

Source: Survey result (2011)

4.2 Seedling need assessment of the study areas

Before raising seed sowing on the beds of seedling nursery, conducting the need assessment of seedlings of pilot area is very important point in nursery works. As shown in the table below, the survey carried out in four study area and by compiling the need data of three consecutive years of selected worada's. As it shown in the table below, each Woreda has been increasing their needs of both tree seedlings and coffee seedling. As a survey result in year 2017 Damot woyde was request 20,000tree seedlings and none of coffee seedlings, again it needs in 2017/2018/ about 312500 tree seedling and request 7500 coffee seedling. On the other hand Humbo Woreda provided seedling request in 2017/2018 which was 312500 trees seedlings and 7500 of coffee seedlings where as soddo town was need tree seedlings similar of that of the other worada's which was 312500 and 7500 coffee seedlings. On the same way the needs of these worada's seedling has dramatically increased from year to year. On the other hand, Humbo, Woyde and Soddo zuriya

13

wants to accept greater than500, 000 tree seedlings for their watershed area. The demand of seedlings increase time to time but there is no enough and standard seedling nursery in the area. As the survey result revealed as, most of the study worada's and towns needs millions tree seedling to their watersheds. This demand may be initiated as to maintain and establish permanent nursery in this selected worada's for the sake of easily supply of seedlings for different pilot area near to site. See table (1) below

Table 1 Seedling need assessment the selected worada's

N o	Pilot Worada's	Total	Three years of multi-purpose seedling need assessment data						
			2016/2017		2017/2018			2018/2019	
		Tree seedlings Request	Coffee seedlings Request	Tree seedlings Request	Coffee seedlings Request	Tree seedlings request	Coffee seedlings Request	Total
1	Damot Woyde	20,000	0	312500	7500	6132635	0	6,472,635
2	Humbo	0	0	312500	7500	500,000	0	820000
3	Soddo Zuria	0	0	312500	7500	326000	30,000	676000
4	Soddo Town	0	0	312500	7500	33,500	50,000	403500
	Total	20,000	0	1250000	22500	6992135	80000	8,364,635

Source: Survey result,(2011)

4.3. Community perception towards nursery establishment

Individuals may have deferent attitudes towards natural resource management in general and in seedling nursery establishment and management particularly. Similarly, in the study area, respondents about 93% replied positive idea about nursery establishment which implies they understand its multipurpose benefits to their natural resource conservation. Also the community in the area, reported as, they have a great need of nursery establishment in their village. On the other hand, the other respondents responded as opposition in nursery establishment in their surrounding area. Even if about 7% of households were responded as negative answered in nursery establishment, the reason behind it is they have no surplus lands to construct nursery in their locality other than the farm land. See figure below (5).

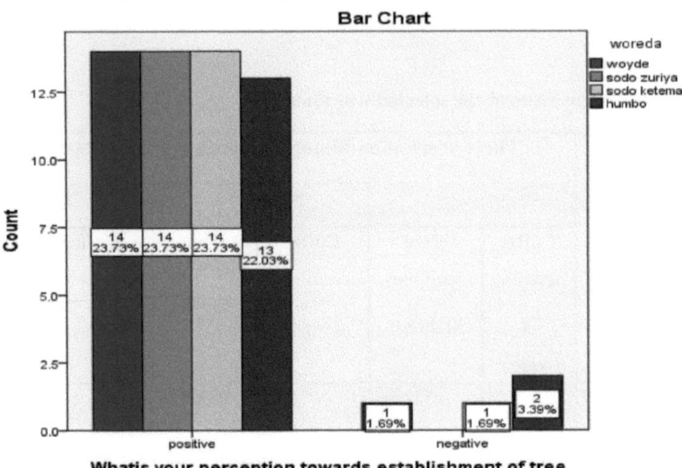

Figure 6 Community perception towards seedling nursery establishment

 Source: Survey result (2011)

Likewise, the survey result shows the respondents rate on the seedling nursery establishment. In this rating, about equally15% from each worada's respondent's response as, they have an interest nursery establishment in their vicinity highly. That means averagely 60% from four worada's respondents were replied as "very high", about 24% were responded as "low" where as16% were replied as "High". The survey result, revealed as the farmers in the study area have an interest on seedling nursery establishment in their surroundings See figure (6) below.

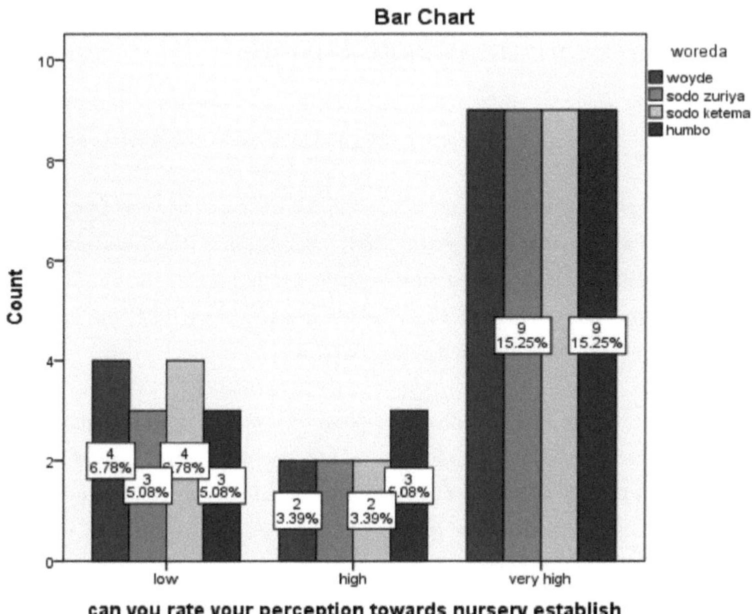

Figure 7 the rates of respondents on nursery establishment

Source: Survey result,(2011)

CHAPTER FIVE

5. Conclusion and Recommendation

Ethiopia is one of the countries where forest resources degradation exhibits primarily due to population explosion. However, the forest policy is not said to be effective in forest policy regards.

Multipurpose seedlings play an important role in the socio-economic development of communities, as it planted on farms provide timber, fuel wood, fodder, fruits, medicine, windbreaks, and a whole range of other economic and environmental benefits. At present the need to plant trees on farms is on the increase. It is difficult, however, for smallholders to access at the right time, in the right quantities and of high quality the trees that they want to plant.

As survey result indicated, the needs of different seedlings by community increased time to time, but yearly supply of seedlings to the designated area has very poor. As a survey result in year 2017 Damot woyde was request 20,000tree seedlings and none of coffee seedlings, again it needs in 2017/2018/ about 312500 tree seedling and request 7500 coffee seedling.

On the other hand Humbo Woreda provided seedling request in 2017/2018 which was 312500 trees seedlings and 7500 of coffee seedlings whereas soddo town was need tree seedlings similar of that of the other worada's which was 312500 and 7500 coffee seedlings. On the same way the needs of these worada's seedling has dramatically increased from year to year. On the other hand, Humbo, Woyde and Soddo zuriya wants to accept greater than500, 000 tree seedlings for their watershed area.

In similar way, the perception of community towards nursery establishment in their area were highly positive. Most of in the study area, farmers have a great interest on nursery establishment and they need financial and skill support from the zonal government.

5.1 Recommendation

➢ Most of in this study areas, there is government nursery sites: but it has been performed with poor management and constraint with lack of enough budget allocation. So it should be given strong attention by any concerned body.

➢ The community should be aware about benefits of trees for their designated areas. By providing trainings to farmers and community related with nursery establishment, manage ment and plantation of seedlings for planned areas are critical activities, so it has to be conducted.

➢ Now a day our country forests has declined, due to human and natural factors; from it the human intervention take a lion share. So any government, nongovernmental sector and community should put their positive foot blue print in conserving   natural and plantation forests.

➢ Special and achievable policies, regulations and legislative measures should be designed and implemented. In spite of there is a forest policy and regulations in the country, there is still a wide gap in implementation.

6. References

Joel Buyinza and Vincent (2016).Improving Sustainable Productivity in Farming Systems and Enhanced Livelihoods through Adoption of Evergreen Agriculture in Eastern Africa Reference.

Teshome (2014).Governing commons and natural resource management, Bahir Dar University, Ethiopia.

Tedla.S and Lema.K (1998). ENVIRONMENTAL MANAGEMENT IN ETHIOPIA:Hav the National Conservation Plans Worked? Environmental Forum Publications Series No.1, OSSREA, Addis Ababa, Ethiopia pp7.

Giliba,R.A.,Boon,E.k.,Kayumbo,C.J.,Chirenge,L.I.,and Musamba,E.B.(2011). The Influence of Socio-economic Factors on Deforestation: A Case Study of the Bereku Forest Reserve in Tanzania. Forest Training Institute, Arusha, Tanzania, PP32

Otu et al.(2011). Impact of Population Growth on Forest Resource Degradation in Ikom Local Government Area. *Department of Geography and Regional Planning, University of Calabar-Nigeria,*

Mowo JG. 2013. Nursery management, tree propagation and marketing: A training manual for smallholder farmers and nursery operators. Nairobi: World Agroforestry Centre

Holden, S.T. and P.L. Sankhayan, 1996, *Population Pressure, Agricultural Develop-ment and Environmental Degradation − A Study in the Himalayan Region.* Project proposal submitted for funding to the Research Council of Norway. Ås: Depart-ment of Economics and Social Sciences, The Agricultural University of Norway.

Debel et al (2014). The Impact of Population Growth on Forestry Development in East Wollega Zone: The Case of Haro Limu District. College of Social Science and Education, Wollega University, Nekemte, Ethiopia pp 90

Denscombe (2007).The Good Research Guide: For Small-Scale Social Research Projects

Cavaye (1996).Qualitative case studies in operations management: Trends, research outcomes, and future research implications